These books are dedicated to Emily, who inspired the author.
—Susan K. Mitchell

Published in 2020 by Enslow Publishing, LLC.
101 W. 23rd Street, Suite 240, New York, NY 10011

Copyright © 2020 by Enslow Publishing, LLC.

All rights reserved.

No part of this book may be reproduced by any means without the written permission of the publisher.

Library of Congress Cataloging-in-Publication Data

Names: Hurt, Avery Elizabeth, author. | Mitchell, Susan K., author.
Title: Prickly protection : stingers, barbs, and quills / Avery Elizabeth Hurt and Susan K. Mitchell.
Description: New York : Enslow Publishing, 2020. | Series: Animal defense! | Audience: Grade 3–6. | Includes bibliographical references and index.
Identifiers: LCCN 2018050379| ISBN 9781978507173 (library bound) | ISBN 9781978508156 (paperback)
Subjects: LCSH: Animal defenses—Juvenile literature.
Classification: LCC QL759 .H84 2020 | DDC 591.47—dc23
LC record available at https://lccn.loc.gov/2018050379

Printed in the United States of America

To Our Readers: We have done our best to make sure all website addresses in this book were active and appropriate when we went to press. However, the author and the publisher have no control over and assume no liability for the material available on those websites or on any websites they may link to. Any comments or suggestions can be sent by email to customerservice@enslow.com.

Portions of this book originally appeared in *Animals with Wicked Weapons: Stingers, Barbs, and Quills*.

Photo Credits: Cover, p. 1 (porcupine) Seregraff/Shutterstock.com; cover, p. 1 (bees) Aedka Studio/Shutterstock.com; p. 6 Photo Researchers/Science Source/Getty Images; p. 8 Amy Nichole Harris/Shutterstock.com; p. 11 Danita Delimont/Gallo Images/Getty Images; p. 13 Darby Jones/Shutterstock.com; p. 16 Jean-Franois Noblet/Biosphoto/Getty Images; p. 18 MollyAnne/iStock/Getty Images Plus/Getty Images; p. 21 Coatesy/Shutterstock.com; p. 24 Audrey SniderBell/Shutterstock.com; p. 27 Francesco Tomasinelli/Science Source; p. 30 Robert Briggs/Shutterstock.com; p. 32 HotFlash/Shutterstock.com; p. 35 scubaluna/iStock/Getty Images Plus/Getty Images; p. 37 Arics Sutanto/Shutterstock.com; p. 40 aaron heritage/Shutterstock.com; p. 45 John Cancalosi/Photolibrary/Getty Images.

Contents

Introduction 4

1 The Fight to Survive 7

2 Coming to the Point 15

3 A Hairy Defense 23

4 Defense in the Deep 31

5 Slow and Spiny 39

Glossary 46

Further Reading 47

Index 48

Introduction

For most animals, life in the wild is hard. It is also dangerous. Animals have to find places to live and raise their families. They have to find enough food to eat. Sometimes, they have to keep from being food for other animals!

Some animals are **predators**. This means that they hunt and eat other animals. Lions, tigers, and wolves are predators. Some animals are **prey**; they are hunted and eaten by other animals. Rabbits and deer are prey. Many animals are both predator and prey. They eat one kind of animal. Another kind of animal eats them. Frogs eat moths. Snakes eat frogs. Owls eat snakes. It's an animal-eat-animal world out there!

Introduction

Animals have to protect themselves from enemies trying to eat them. They also have to protect their young. Even top predators are at risk of being eaten when they are babies.

Sometimes, animals have to protect their homes from other animals. Sometimes, they have to fight for a mate. There are many times in an animal's life when it has to fight to protect itself.

To survive, animals must have a few tricks. And they do. Each kind of animal has **adaptations,** or special skills or features that help them find food. They have special skills that help them build homes or nests for raising their young. Animals that hunt other animals have many hunting tricks. Some animals have tricks for hiding themselves from their enemies. All animals have unique adaptations that help them survive in the place where they live. These adaptationss are different for each kind of animal.

The survival skills animals use to protect themselves are called **defenses**. These are built-in weapons that animals can use when they are attacked, such as quills, **barbs**, and stingers.

PRICKLY PROTECTION: Stingers, Barbs, and Quills

Bees have sharp stingers for protection. Other animals run away to keep from getting stung by bees.

When people poke themselves with sharp needles or get stung by bees, it can hurt a lot. It can hurt so much that they forget what they were doing to take care of their fingers.

When an animal gets stung or poked with something sharp, it does the same thing. If it is attacking another animal, it might just stop. This gives the other animal time to get away. That's why stingers and **prickles** are good defenses for animals.

Chapter 1
The Fight to Survive

Animals have many different ways of staying safe. Some animals use **camouflage** to blend in with their surroundings. The animals that are trying to eat them can't see them. Some animals look like more dangerous animals so that predators will avoid them. Some animals, such as rabbits, can outrun their predators.

There are times, however, when an animal has no choice but to stand its ground and fight. If an animal is going to fight, it needs the right weapons. It may have sharp, spiky skin or hair. It may have long horns or tough hooves. No matter how great the weapon, animals use them only if they have to.

PRICKLY PROTECTION: Stingers, Barbs, and Quills

A water buffalo may not seem like a match for a lion. But those sharp horns and flying hooves can drive lions away.

Animals will often try to frighten off the predator before fighting. They will make a lot of noise. They may show their weapons to the predator. Sometimes, if a predator sees that an animal is able to fight back, it will not attack. These threats do not always work. That is when an animal might be forced to use its special weapon. Using weapons for defense, however, can be risky.

The Fight to Survive

One problem is that many predators have weapons of their own. They may be armed with sharp teeth and claws. To use a weapon for defense, an animal must be very close to its predator, which is obviously not safe.

Many animals have special weapons to help with these problems. One of these special defenses is a prickly body. Spikes or spines are special hard hairs or skin that stick up all over an animal's body. When the predator tries to eat the animal, it gets a mouthful of pain!

Not Always Defense

Prey animals use their weapons to protect themselves from predators. Predators have weapons, too. They use these to catch and kill prey.

Animals with sharp claws can slash at their prey. Strong jaws and sharp teeth help in their attacks. Predators are often large and can move fast. Many predators' weapons deliver poison, such as a **venomous** snake's sharp fangs. Some fangs are powerful and sharp enough to pierce the toughest skin. A snake's bite and the venom it delivers can be a deadly combination.

PRICKLY PROTECTION: Stingers, Barbs, and Quills

Some prey animals use poison as a defense. Male platypuses have poisonous **spurs** on their back legs. They mostly use these weapons to fight other male platypuses. The poisonous spurs can also be used for defense against predators.

Other animals use their feet and legs as weapons. The kangaroo and zebra are two animals that will kick to defend themselves. Few animals, however, pack as big a punch for their size as the weta. A weta is a huge insect that lives in New Zealand. It looks very much like a giant grasshopper. When threatened, the tree weta will kick

Bone Breaker

Animal defenses usually harm the attacker. But sometimes, the animal defending itself gets damaged, too. The hairy frog breaks its own bones! When an enemy attacks, this little African frog squeezes the muscles in its back leg, breaking the bones. Then it pushes the bones all the way through the skin, making a sharp claw. The frog uses the claw to fight off the attacker. Fortunately, frogs heal quickly.

The Fight to Survive

with its long back legs. The kick itself does not hurt. The many spines along its back legs do, however. They can cause a nasty scratch to the predator that could become infected. It's best not to bother a weta!

There is one big difference between predators' weapons and those used only for defense. Predators' weapons are meant to kill. On the other hand, weapons

A ram is a male sheep. These rams butt each other with their big horns. They fight for territory and for females.

PRICKLY PROTECTION: Stingers, Barbs, and Quills

for protection are usually designed to cause injury or pain—not death. Sometimes, they just confuse or irritate the predator. Their purpose is to give an animal in danger some time to get away.

Eating and being eaten aren't the only times animals use weapons. Next to eating, having babies is one of the most important things in any animal's life. Many male animals use their weapons to fight for females. They use horns, hooves, tusks, or teeth. It is rarely a fight to the death, though. Usually one male simply gives up and goes away. The winner is the male that best uses his weapons. He is the one that gets the female.

Animals also fight each other for **territory**. Having enough room to live, eat, and raise young is very important to an animal's survival. Any animal that comes near the home or hunting grounds of another animal better be ready for a big fight.

Stay Away from My Babies!

Once animals have mated and started a family, they use their weapons to defend their young. Young animals

The Fight to Survive

This cheetah cub is too young to protect itself. Its mother hides it in the grass so that predators can't easily see it.

PRICKLY PROTECTION: Stingers, Barbs, and Quills

Fun Fact!

When threatened, opossums pretend to be dead. Most predators like to kill their own food. They won't bother with an animal that they think is already dead.

usually aren't able to protect themselves. Their weapons may not be fully developed, or they might not know how to use them until they are older. Some babies rely on their parents to protect them. An animal may hide its young from predators. But sometimes, animals have to fight to protect their young. Predators and prey will fight to the death to protect their babies.

Chapter 2
Coming to the Point

Porcupines are famous for their weapons. Thousands of sharp quills cover a porcupine's body. They look dangerous, and they are! An animal that tries to attack a porcupine can come away with some very painful wounds.

The quills of all porcupines are just special hairs. When a baby porcupine is born, its quills are very soft. But it takes less than an hour for the baby's quills to harden and become weapons. The quills grow just like any other hair. They keep growing throughout the porcupine's life.

When a porcupine's quills get too long, they break off at the base. Sometimes, they break off during a battle with another animal. There are always new quills growing to replace any that are lost.

PRICKLY PROTECTION: Stingers, Barbs, and Quills

The quills of this African crested porcupine can cause a lot of damage to a predator.

Duck and Cover

Porcupines are very shy animals. They are usually nocturnal. This means they are active mostly at night, and they sleep during the day. Porcupines spend a lot of time in trees. They love to eat twigs and bark. They have large front teeth that help them chew on wood.

Coming to the Point

When a porcupine is out in the open and looking for food, it is in the most danger of being attacked. Normally, a porcupine's quills lie flat against its body. When threatened, special muscles in the porcupine's skin raise all of the quills up straight. But a porcupine does not use its quills right away. It usually tries to scare a predator off first by making all kinds of noise.

One kind of porcupine is especially good at this. It has special quills on its tail that are hollow. When it shakes these quills, they knock against each other, making a rattling sound. It's loud enough to scare off some predators.

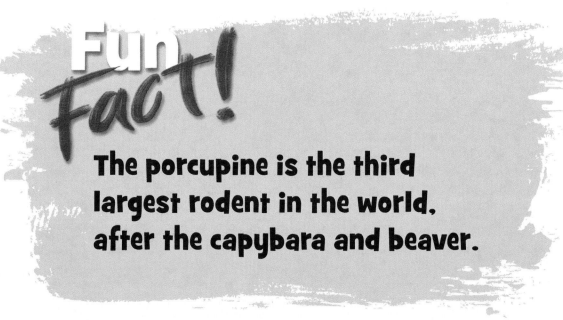

Fun Fact!

The porcupine is the third largest rodent in the world, after the capybara and beaver.

PRICKLY PROTECTION: Stingers, Barbs, and Quills

This poor dog met a porcupine up close. His nose will be sore for a long time.

Coming to the Point

If that doesn't work, then the porcupine turns its back to the enemy. Porcupine quills always point backward. By turning around, the porcupine prepares its weapons. It also protects its head from an attack.

Sometimes, a porcupine will scoot backward toward its enemy. If the animal doesn't run away, it gets a face full of sharp quills. The quills stick inside the animal's skin and break off from the porcupine. It distracts and stuns the predator. This gives the porcupine enough time to get away from its attacker.

Even after the porcupine leaves, the attacker is still in trouble. Tiny barbs cover the surface around the end of

Clean that Wound

Even if the predator kills the prey, it might get wounded. Injuries can be dangerous. Wounds often become infected. Minor wounds heal quickly, but big ones can be deadly. When the prey shows its defense, the predator has to think fast: "Will I be injured if I fight this animal?" The safest choice may be to find an easier supper.

PRICKLY PROTECTION: Stingers, Barbs, and Quills

each quill. These barbs face in the opposite direction from the quill point. This makes the quills very hard to remove. When an animal pulls at the quill stuck in its skin, the barbs push the opposite way, causing even more pain.

Over Easy

A prickly coat does not mean the porcupine is out of danger. In Africa, lions and hyenas love to eat porcupines. In North America, a weasel-like animal called the fisher has porcupine on its menu. These animals have figured out the porcupine's one weak spot.

There are no quills on a porcupine's belly. It is soft, furry, and completely unprotected. Many predators have learned to flip porcupines over to attack them. It is not as easy as it may seem. Some animals walk away with a few quills stuck in them even if they win the fight!

Not Just Porcupines

Porcupines aren't the only animals with quills. Hedgehogs have quills, too. But hedgehog spines don't have barbs. Also, hedgehog spines don't fall out very easily.

Coming to the Point

This European hedgehog has quills, too. These quills aren't as dangerous as a porcupine's quills, but they're still good protection.

PRICKLY PROTECTION: Stingers, Barbs, and Quills

Hedgehogs don't have spines on their faces, legs, or bellies. But they have a trick to protect these parts. They can roll up into a tight ball. When they do this, they tuck their soft parts inside. An animal trying to eat a hedgehog will get a mouthful of quills.

Hedgehogs are smaller than porcupines. They also have different tastes in food. Porcupines prefer plants. Hedgehogs like to eat worms, snails, and even small snakes. Hedgehogs don't climb trees. They build nests on the ground. Hedgehogs live in Europe, Asia, and Africa. There are no wild hedgehogs in North or South America.

The spiny anteater is another animal with spines. It has a long sticky tongue to help it slurp up ants. Its quills protect it from its enemies. Spiny anteaters look like porcupines and hedgehogs, but they aren't related to them. The spiny anteater is also called an echidna. It lives in Australia, Tasmania, and New Guinea.

Chapter 3
A Hairy Defense

If someone is looking for a spider to star in a horror movie, the tarantula is their best bet. Plenty of humans are very afraid of them. Some types of tarantula can have a 12-inch (30½-centimeter) leg span. All tarantulas have huge fangs. Tarantulas don't use the nasty-looking fangs for defense. Instead, they use their fangs to get food. Tarantulas have another surprising weapon to use for defense: hair. The tarantula's body is covered with thousands of tiny hairs. When the tarantula is threatened, it throws that hair at its attacker.

Hair Me Out

Most tarantulas live in underground burrows. They live in many parts of the world. They are active mostly at night.

PRICKLY PROTECTION: Stingers, Barbs, and Quills

This is a Chilean rose hair tarantula. Its big, hairy body and long legs make it look scary, but it's not dangerous to humans.

They don't move fast, but they are good hunters. They prey mostly on insects. Sometimes, they eat bigger **game**, such as frogs, toads, and mice.

A tarantula's hairs are very important—and not just for defense. The hairs help the spider feel vibrations. When another animal walks, hops, or flies nearby, it

A Hairy Defense

causes small vibrations. The hairs on the spider's body feel these movements. This lets the spider know exactly where lunch might be.

When a tarantula feels food wandering nearby, it rises up on its back legs. Then the spider pounces on the animal with its two huge fangs. The fangs inject venom into the prey. This does two things. First, the venom **paralyzes** the animal so it cannot move. Next, the venom turns the animal's insides into liquid. Tarantulas cannot chew. They have to drink their meals. This sounds scary and gross, but no tarantulas are actually dangerous to humans. Their bite feels more like a wasp sting. In fact, many people keep tarantulas as harmless pets—fangs and all!

Tarantulas don't have many natural predators. But some animals do love to have a nice lunch of tarantula. Weasels, skunks, snakes, owls, and other large birds sometimes catch and eat tarantulas. They might look elsewhere for a meal, though. The big hairy spider has a dangerous weapon to protect itself. When the tarantula feels threatened, it throws its hairs at its enemy. Like most animals, the tarantula does not fight

PRICKLY PROTECTION: Stingers, Barbs, and Quills

at first. It often stands up tall on its back legs. It shows its fangs to its enemy. The spider hopes this bluff will work to scare the other animal away.

When it doesn't, the tarantula gets serious about defense. It whips out a ninja move. It turns around and backs toward the attacker. It uses its rear legs to

What Web?

Most spiders spin webs to catch their prey. Tarantulas do not. Instead, they lie in wait for their prey. When their prey comes close, they pounce. They grab the prey with their legs and inject it with venom. Then they finish it off with their fangs.

Tarantulas do spin silk, though. They use the silk to line their underground burrows. This helps keep out dirt and dust. Sometimes, they make a burrow out of silk. They may even stretch a thin line of silk across the entrance to their burrows. If an enemy comes too close, it will trip on the silk. That lets the tarantula know it has company.

A Hairy Defense

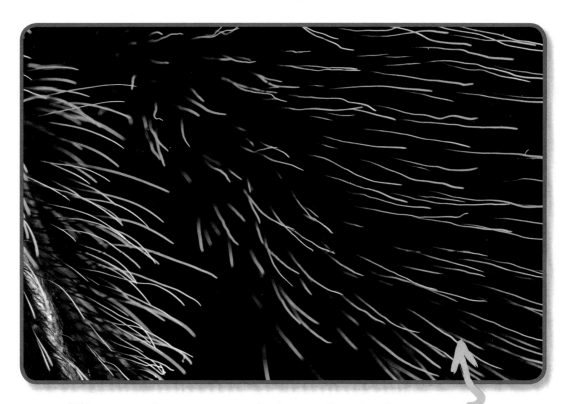

This is what tarantula hairs look like up close. These irritating, barbed hairs are called urticating hairs.

fling hairs off its back. That might not sound too threatening, but these tiny hairs can cause quite a bit of pain. Each little hair is covered with even tinier barbs on the end. These barbs help the hair stick in the enemy's skin. The tiny weapons can cause bad itching and burning for the unfortunate animal that

PRICKLY PROTECTION: Stingers, Barbs, and Quills

Fun Fact!

The biggest spider in the world is the goliath birdeater tarantula. It can be more than a foot (30½ cm) long from leg to leg! Like the name says, it sometimes eats birds.

encounters them. Sometimes, the itching and burning can last for days. The prickly hairs not only sting; they can make attackers blind for a few minutes. This gives the tarantula plenty of time to run to safety.

Stinging hairs are tarantula's defensive weapon, but these spiders have another trick that can sometimes save their lives. If they lose a leg, they can grow a new one. If a predator grabs a tarantula by one of the spider's eight legs, it might not get the whole spider.

A Hairy Defense

Run for Your Life!

Tarantulas have one enemy that is even scarier than they are. The tarantula hawk hunts tarantulas. A tarantula hawk is not a bird. It is a wasp. It is much more frightening than a tarantula. It has the most painful sting of any insect. When a tarantula hawk attacks a tarantula, it injects a powerful venom into the spider. This makes the tarantula unable to move. The wasp then drags the unmoving but still living spider into a burrow. It lays an egg in the spider's body. Then it covers up the burrow with dirt. As the baby wasp hatches and grows, it eats the tarantula. It does this very slowly. It eats the least important parts first. It saves the brain and heart and other important organs until last. That way, the spider stays alive longer. By the time the baby wasp is old enough to go out on its own, the tarantula is finally dead.

Many humans scream and run when they see tarantulas. Tarantulas run for their lives when they see this wasp. Their stinging hairs don't do much to stop this wicked predator.

PRICKLY PROTECTION: Stingers, Barbs, and Quills

The tarantula hawk is beautiful but deadly. No one likes to have a close encounter with this insect, especially tarantulas.

Tarantulas usually fling their hairs during a battle with a predator. But they also use their stinging hairs for a different kind of defense. They sometimes place the hairs around the entrance to their burrows. This is a way of marking territory. It lets other tarantulas know who lives there. Marking territory can keep males from having to fight other males. Each animal marks out his own area. But the markings aren't meant only for other spiders. They also let would-be predators know to stay away.

30

Chapter 4

Defense in the Deep

When seen from a shore, the ocean looks like a peaceful place. But there is a lot of excitement going on under the waves. Just like land animals, sea creatures have to protect themselves from predators. They often have to fight for mates, and they have to protect their young.

The ocean can be especially dangerous for slow-moving animals. Sea urchins don't move fast like sharks or eels. They can't get away quickly. But they have another form of defense. Their bodies are completely covered with sharp spikes. This makes sea urchins look like floating pincushions! Sea urchins are sometimes called the porcupines of the sea.

PRICKLY PROTECTION: Stingers, Barbs, and Quills

Sea urchins are in the same animal group as starfish. They are both called echinoderms (a-KINE-a-derms). That is a Greek word meaning "spiny skinned." Sea urchins are the spiniest of all echinoderms. They are protected with an arsenal of spikes. Sometimes, the

This sea urchin has pretty purple spikes and lives on a coral reef in the Indian Ocean.

Defense in the Deep

Fun Fact!

The red sea urchin is one of the longest-living animals on Earth. It may live for as long as two hundred years.

spikes are even poisonous. They are attached to a hard, chalky covering called a test. The test covers and protects the round, soft body of the sea urchin.

Most sea urchins live in salty, shallow waters. They can live in oceans all over the world. Some even live in deeper waters. They eat mostly algae, which are small, plant-like organisms that live in the water. Sea urchins also eat tiny bits of plants or animals that they find on the ocean floor. Sometimes, when a sea urchin is looking for food, it finds itself on the menu. Many animals think sea urchins are a delicious meal!

PRICKLY PROTECTION: Stingers, Barbs, and Quills

Sea otters like to dive for sea urchins. When they find one, the otters smash them open for a tasty snack. The sea urchin is not even safe from its cousin, the starfish. But predators do not just come from the ocean. The sea urchin has enemies in the air also—many ocean birds eat sea urchins.

The soft body parts inside the sea urchin's tough outer covering is what animals want to eat. An animal that wants sea urchin for a meal has to do battle with the urchin's spiny body. Animals planning on a lunch of sea urchin need to be careful.

The Backup Plan

There are more than nine hundred different kinds of sea urchin. They come in many different colors and sizes. Their spines can be very different also. Most sea urchins have sharp, pointed spines. They can be long, short, or in between. These sharp spines often break off when the sea urchin is attacked. They can stick in the skin of a predator and cause a nasty wound. Many of the pointy sea urchins have spines that are poisonous.

Defense in the Deep

Up close, this sea urchin's spines don't look very sharp. But they can cause a lot of damage and pain.

Sea urchins don't depend just on their spines for protection. They also have a backup defense. In between their spines are tiny claws. Sea urchins use these tiny pincers to clean themselves. They also help find and grab food. But the tiny pincers can also be poisonous just like the spines. If some animal manages to defeat the sea urchin's spines, the pincers are there as a second defense.

PRICKLY PROTECTION: Stingers, Barbs, and Quills

Poison Puff Ball

Sea urchins aren't the only sea animals that use spines and spikes for defense. Some fish use spines as weapons. The pufferfish can change its shape when threatened. Normally, it swims along and looks like any other fish. Its spines lie flat against its side. These spines are really

Sea Urchins to the Rescue

Sea urchins are good at protecting themselves. They may be good at protecting everyone else on Earth, too. Carbon dioxide is a chemical released when humans burn fuels to run cars and power plants. Too much carbon dioxide is making the planet too warm. Scientists in England discovered urchins take carbon dioxide from seawater and turn it into a harmless chalk. They want to learn how seas urchins do this so that they can try to lower the amount of carbon dioxide in the sea and help slow global warming.

Defense in the Deep

This pufferfish has turned itself into a spiky ball. It will be hard for a predator to eat it now.

special scales on the fish's body. When a predator attacks a pufferfish, the pufferfish gulps water into its body. As its body becomes round, its spines pop up. They stick out in all directions. The pufferfish looks like a big, spiky ball!

This helps the pufferfish in two ways. First, puffing up makes it too big to swallow. Second, the spiky skin can hurt the predator.

Some types of pufferfish, such as the fugu, have more deadly defenses. The fugu has very poisonous internal organs. Eating it would be deadly for any animal. There is enough poison in one fugu to kill thirty people. In Japan, these pufferfish are a treat for some people! They know

PRICKLY PROTECTION: Stingers, Barbs, and Quills

that eating them can be dangerous. A specially trained chef prepares the fish to get rid of the poison. But even so, nearly fifty people die in Japan each year from eating fugu.

Watch That Tail!

The stingray is another sea animal that uses spikes for protection. They carry their weapons on their tails. These animals are related to sharks. Their skeletons are not made of bone. Instead, they are made of cartilage. This is the same bendable material that human ears are made of.

Stingrays are usually peaceful creatures. They glide along the ocean floor looking for food. But when they sense a threat, they launch their weapons. The stingray whips its tail around to stab the predator with the sharp spike. Near the tail spike are venom **glands**. These release poison along with the stab from the stingray's tail. The hard, sharp spike itself can cause a nasty wound. Then the poison causes a painful sting. These two things together are enough to give the stingray a chance to get away from a predator.

Chapter 5
Slow and Spiny

Some animals depend on one type of defense to keep them safe. But some animals have lots of weapons in their arsenals. When it comes to defenses, the thorny devil has it all. This spiny Australian lizard uses camouflage to blend in with its desert environment. It can change colors depending on the type of ground it is on to blend in even better. It also has a "false head" that tricks predators into attacking that instead of the thorny devil's real head. Next, the thorny devil can puff up to look large and scary. As if all that weren't enough, as its name suggests, it is covered in thorns!

Despite its scary name, the thorny devil is not very big. It is only about 6 inches (15 cm) long. Female thorny devils lay between three and ten eggs each year.

PRICKLY PROTECTION: Stingers, Barbs, and Quills

The thorny devil has thorns all up and down its body. It also blends in with the red desert sand.

The eggs hatch about three to four months later. The babies are already fully armed with spikes and ready to be on their own when they hatch.

During warm days, this prickly lizard hunts for ants. Thorny devils are fast eaters. They have been known to eat up to forty-five ants a minute!

Slow and Spiny

But this is the only thing about the thorny devil that is fast. It is a very slow-moving lizard. Because it is so slow, it needs all of its defenses for protection. Many animals try to eat the thorny devil despite its spiky skin. Large birds called bustards and even lizards called goannas hunt for thorny devils.

Fun Fact!

The thorny devil is sometimes called a moloch. Moloch is a terrible demon character in a famous poem, *Paradise Lost*, by John Milton (1608–1674). Scientists named the lizard moloch after Milton's demon.

PRICKLY PROTECTION: Stingers, Barbs, and Quills

Standing Its Ground

Unlike many animals, the thorny devil does not run away when threatened. It is just not fast enough to escape. Instead, the thorny devil stands its ground. It relies on its many defenses for protection. Changing color is one trick. The thorny devil's brownish color helps it blend in with the sandy Australian deserts. But it can also change to different shades of brown, red, and yellow to blend in with whatever color sand it happens to be walking on at the time.

If that does not work, and the thorny devil is spotted by a predator, it puffs itself up with air. This makes it seem bigger and more dangerous than it really is. This defense bluff sometimes works and scares off a predator. But not always. Next on the list of the thorny devil's defenses is its false head.

On the back of the thorny devil's spiky neck is a large bulge also covered in spikes. The thorny devil ducks its head and lifts this fatty area. Animals might think this is the thorny devil's head. They will be tricked into attacking this area. The thorny devil's real head stays protected. It might be able to get away.

Slow and Spiny

What the thorny devil really counts on, however, are the sharp and pointy spikes on its skin. These spikes are not poisonous. But they do make the thorny devil very hard to eat.

The pointy spikes of the thorny devil also help it get water. The deserts of Australia are very dry. Little rain falls there. This means that it is very hard for animals to find drinking water. The thorny devil's spikes form

Learning from Animals

Police sometimes use guns to protect people from criminals. They try to use non-deadly weapons if they can. They've learned a few tricks from animals.

One non-deadly weapon police use is a Taser. Tasers are electric stun guns. Tasers deliver an electric charge, much like the one that electric eels use. A Taser gives just enough of a jolt to stop a person, but not enough to kill. Still, Tasers can be dangerous and should never be handled by anyone not trained to use them.

PRICKLY PROTECTION: Stingers, Barbs, and Quills

a system of grooves along the lizard's back. Any water that falls on the thorny devil's back travels down these grooves. The grooves lead down to the thorny devil's mouth for a welcome drink.

Seeing Blood!

Thorny devils aren't the only lizards that have horns. But for some animals, they are not the primary defense. The horned lizard is a distant cousin of the thorny devil. Horned lizards are found in the southwestern United States and in parts of Mexico. Some types of horned lizard also have one of the weirdest defenses of any animal in the world. When attacked, horned lizards shoot blood out of their eyes!

When it senses danger, the horned lizard's brain signals blood vessels near its eyes to break. Special **ducts** let some horned lizards shoot up to one-third of the blood in their body. They can shoot it more than 3 feet (1 meter) at a predator.

This weird weapon confuses and frightens a predator. That gives the horned lizard enough time to get away.

Slow and Spiny

That's real blood coming out of this horned lizard's eye. But it's not because it's hurt. It's a trick so that the lizard doesn't get hurt.

As with all creatures, the horned lizard's body is always making new blood cells. The lost blood will slowly be replaced.

Most animals will always choose to flee danger when they can. Those with horns, hooves, spines, spikes, quills, or even poison can stand and fight if they have to. Animals with weapons at least have a fighting chance against predators.

Glossary

adaptation A special skill or physical feature that an animal has developed over time to help it survive in its specific environment.

barb A sharp point that faces in the opposite direction from a main point.

camouflage A defense in which an animal's coloring or shape helps it hide from predators.

defense Protection against an attack.

duct A tube or opening in the body.

game Animals that are hunted and eaten by other animals.

gland An organ in an animal's body that releases chemicals.

paralyze To make unable to move.

predator An animal that hunts and eats other animals.

prey Any animal that is a food source for other animals.

prickle A sharp object that sticks out from a plant or animal, such as a spine or thorn.

spur A sharp, clawlike object that is filled with poison on the back leg of a platypus.

territory An area of land claimed by an animal and defended against other animals of the same species.

venomous Producing a substance that becomes toxic when injected into a victim, usually through biting.

Further Reading

Books

Avery, Sebastian. *Porcupines: Creatures of the Forest Habitat*. New York, NY: PowerKids Press, 2017.

Fletcher, Patricia. *Why Do Thorny Devils Have Two Heads? And Other Curious Reptile Adaptations.* New York, NY: Gareth Stevens, 2017.

Johnson, Rebecca L. *When Lunch Fights Back: Wickedly Clever Animal Defenses*. Minneapolis, MN: Millbrook Press, 2015.

West, David. *Spiders and Other Creepy-Crawlies*. New York, NY: Windmill Books, 2018.

Websites

Animal Diversity Web
animaldiversity.org/collections/spinesquills
Learn more about animals with spines and quills.

Animal Planet: Top 10 Animal Weapons
www.animalplanet.com/wild-animals/10-animal-weapons
Explore ten different, and often surprising, animal weapons.

San Diego Zoo: Porcupine
animals.sandiegozoo.org/animals/porcupine
Find out more details about the porcupine.

Index

A
anteaters, spiny, 22

B
barbs, 5, 19–20, 27
bees, 6

C
claws, 9, 10, 35

E
echinoderms, 32

F
fangs, 9, 23, 25, 26
frogs, 4, 10, 24

G
glands, 38

H
hedgehogs, 20, 22
hooves, 7, 12, 45
horns, 7, 12, 44, 45

O
opossums, 14
owls, 4, 25

P
platypuses, 10

porcupines, 15–22, 31
pufferfish, 36–38

Q
quills, 5, 15–22, 45

S
snakes, 4, 9, 22, 25
spikes, 9, 30, 31, 32–33, 36, 38, 43, 45
spines, 9, 11, 20, 22, 34–35, 36–37, 45
stingers, 5, 6
stingrays, 38

T
tarantulas, 23–30
Tasers, 43
thorny devils, 39–44
tusks, 12

U
urchins, sea, 31–35, 36

V
venom, 9, 25, 26, 29, 38

W
wasps, 25, 29
wetas, 10–11